FIRE IN YOUR HOME
How to prevent it.
How to survive it.

Copyright © 1978
NATIONAL FIRE PROTECTION ASSOCIATION
All rights reserved

NFPA No. SPP-52
Library of Congress No. 78-60515
ISBN 0-87765-131-0

National Fire Protection Association
Batterymarch Park, Quincy, Massachusetts 02269

Why a book on firesafety?

Because 7,800 Americans died in home fires last year. On the average, 21 people a day die in fires in their homes. This amounts to 80 percent of all fire deaths. Few are marked by burns; over half die from smoke inhalation, often while they sleep.

Tragedy and pain mar the lives of the 300,000 Americans seriously burned each year. A hospital stay of six weeks is average and hospital costs of $60,000 are not unusual.

Nothing—life savings, federal aid, insurance—can come close to replacing the lives lost and possessions destroyed in fires.

By reading this book, you stand to gain three distinct defenses against fire. One, you'll be better able to keep fire out of your home. Two, if fire does strike, your family will have the advantage of an early warning. And, three, you'll have an escape plan that can save your life.

The United States has the worst fire record of any industrialized nation. Hundreds of thousands of fires rage each year, wreaking damage in excess of 6 billion dollars, leaving thousands homeless and jobless.

Read this book.

In eight brief easy-to-read chapters, the nation's top firesafety experts from the National Fire Protection Association tell you what you need to know, and do, to keep you and your family safe from the tragedy of fire.

TABLE OF CONTENTS

CHAPTER 1 ... Fire! 5

CHAPTER 2 ... Buying Time: Home Fire Detection 8

CHAPTER 3 ... Meet EDITH 14

CHAPTER 4 ... Survival 20

CHAPTER 5 ... Specially Tailored Plans 26
Elderly and Handicapped
Young Children
Apartments
Babysitters
Mobile Homes
Pets

CHAPTER 6 ... Learn Not To Burn 30

CHAPTER 7 ... Firesafe Life Style 37

CHAPTER 8 ... When You're On The Spot 42

A FINAL WORD 44

HOME FIRESAFETY KIT 45
Fire Prevention Checklist
Babysitter Checklist
Escape Plan

Research and development of this book was made possible in part
through a grant from the Teledyne Water Pik Corporation.

CHAPTER 1
Fire!

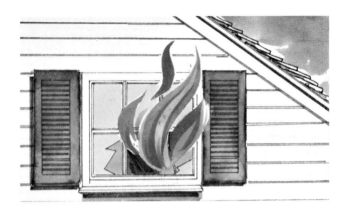

Fire!

In your home? Of course not, you say.

But what if . . .

You'd know what to do, right?

Or do you think, as one man did, that the bathroom shower will protect you from the fire's heat? That you can rush from room to room carrying out your children? That you have the strength of love and body to dash back into the burning house to save a member of your family?

Don't count on it. Until you know the facts, or confront the reality of fire, it's difficult to picture just how you would react.

One basic point you may not know: If fire hits your home, you may never wake up—the smoke and toxic gases work that silently and quickly. Or if you do wake up, and stand up, the same smoke and gases can still knock you out.

Maybe you know to keep low and crawl. But will you open the door without testing it, allowing smoke and superheated air to rush in and possibly kill you?

In a way, fire is a child. To be born and grow, fire needs nurturing. And without thinking, man contributes to tragic fires by supplying the three prime ingredients—oxygen, fuel

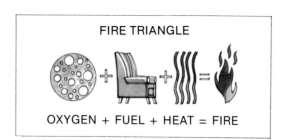

and heat.

Almost any material can become fuel for a fire—clothes, furniture, plastics, flammable liquids, wood. Actually, it's not these objects that cause fire, but their vapors that combine with oxygen and ignite when heated to a high enough temperature.

There's ample oxygen in the air to feed a fire. With sufficient oxygen, it takes only a spark, an open flame or high heat to start a fire.

Here's what happens in a fire. Fire consumes the oxygen, feeding itself, starving you. Normal oxygen content of the air we breathe is about 21 percent. During a fire, the oxygen level rapidly drops. If it falls below 17 percent, clear thinking and muscle control become difficult and your attempts to escape become irrational. For example, you might claw at a door rather than simply turning the knob. When oxygen falls

below 6 percent, breathing stops. After 4 to 6 minutes without oxygen, brain death occurs.

The fire also produces superheated air, smoke and toxic products, any of which can kill you long before the flames ever reach you.

The various toxic gases work in different, yet deadly, ways. Since the fire is taking oxygen you need, your breathing speeds up in an attempt to get more oxygen to your lungs. You also breathe faster because of exertion, heat and fear. This faster breathing means inhaling greater amounts of the poisonous gases. Several of the gases impair sight, smell, even your reasoning power.

Many materials in your home will give off toxic gases when they burn. Some of the toxic products shown to be the main causes of fire deaths include:

carbon monoxide — *hampers oxygen from reaching the brain; the most abundant of fire gases. Invisible and odorless. Produced in all fires.*

carbon dioxide — *overstimulates the rate of breathing; thus increases the intake of other toxic gases.*

hydrogen sulfide — *affects the nervous system; causes dizziness and pain in the respiratory system.*

nitrogen dioxide — *extremely toxic; numbs the throat.*

These and other gases work in such a way as to "fool" your senses. You may never be aware of the deadening effects. You may lose consciousness, perhaps until it is too late.

Fire generates superheated air and gases that cause loss of consciousness or death within several minutes. Our bodies can bear temperatures between 150° and 250°F for only moments. Human tolerance is no match for fire's heat.

Yet, smoke is probably fire's major threat. It moves fast, hampering visibility, obscuring light, and blocking vision. Smoke particles irritate the respiratory system, impair sight and cause coughing and sneezing. The end result: difficulty in breathing, seeing and thinking clearly. All of which can lead to panic.

CHAPTER 2
Buying Time: Home Fire Detection

You can't just rely on your own senses and presence of mind to detect and escape from fire. The very nature of smoke and other fire products may prevent you from realizing the danger and reacting properly.

Most fatal residential fires strike at night, when most people are asleep. To be alerted in time to escape, you need a system that will react, not succumb, to fire.

A home fire detection system can provide you and your family with an early warning—warning that will give you the extra minutes you'll need to escape.

The number of lives saved by smoke detectors is growing every day. A family of six in Mission Viejo, California, was awakened at 5 a.m. on a November morning by their smoke detector. All six made it to safety. An elderly couple in Charlotte, North Carolina, was alerted to a fire by their smoke detector after having been asleep for about three hours. They too got out in time.

In both these and other situations, investigating fire departments have asserted that the smoke detector's early warning saved lives.

The very smoke that can kill you, can save you— by activating the smoke detector.

Hot gases and smoke rise, filling the highest points in a house or room before moving down toward the floor. This property of smoke can be turned to your advantage:

1. A smoke detector, located high on a wall or ceiling where smoke first collects, can sound an early warning even before you smell smoke or see flames.

2. Cleaner air is near the floor. Your best chance of escape is to keep low and crawl on your hands and knees.

Successful escape from a home fire hinges on reaction time and preplanning. The early warning of a fire detector can actually add to your escape time. Make best use of that added time by having a prearranged escape plan for the family.

Installing a smoke detector is not an absolute guarantee of safety. However, it's an important part of your total fire-safety plan, discussed later in this book.

Every member of the family should sleep with the bedroom door closed. You may argue that you want to be able to hear your children, but consider this fact: A closed door can hamper the spread of a fire. So, at least for a few moments the closed door acts as the barrier between you and fire.

The odds of a fire starting in a bedroom are remote, unless you or another family member make a habit of smoking in bed. Do yourself and your family a favor: *Never smoke in bed.*

Basic facts about fire detectors.

Two types of fire detectors are available for home use: heat detectors and smoke detectors. Smoke detectors alone, when properly located and maintained, offer the minimum level of safety recommended by the National Fire Protection Association. Smoke detectors operate on either of two different principles, photoelectric and ionization, in reacting to the products of combustion.

At least two-thirds of the deaths in home fires stem from inhalation of smoke and toxic gases. These deadly combustion products will activate a properly located and maintained smoke detector, awakening the household and allowing time to escape.

Since most home fire deaths occur at night and since smoke rises, it makes sense to install a smoke detector outside the sleeping areas of your home on or near the ceiling.

The best location is in the hallway near your home's bedrooms. Fires generally start in the living areas or in the kitchen. The smoke detector should be in the hallway between these areas and the bedrooms. If yours is a large or multi-level home, you should seriously consider installing two or more smoke detectors.

In NFPA's standard on "Household Fire Warning Equipment," the minimum protection is a smoke detector to protect *each* sleeping area plus one on each additional story of the home.

Guidelines for detector purchase and installation.

Both placement and type of sound must be given great consideration. There are two types of warning sounds for you to choose from: the constant horn-type sound found on most smoke detectors or the siren-type warning sound. This siren sound is the same type used on fire, police and other emergency vehicles.

If heat detectors are used, they should be a part of a total system that includes smoke detectors. Home heat detectors react when air temperature reaches a certain point, usually 135° F. Used in conjunction with smoke detectors, heat detectors are useful in kitchens, attics, basements and attached garages.

A few "do's" when purchasing a smoke detector:

• *Check for the label of a testing laboratory; don't buy a detector that doesn't have a laboratory label.*

- *Read instructions, especially the schedule and means for testing the detector.*
- *Install the detector as recommended by the manufacturer.*
- *Install the detector as soon as possible.*

Consider these locations:

1. Locate your smoke detector near sleeping areas. The preferred location is in hallways or areas adjacent to bedrooms.

2. Protect escape routes. Bedrooms are usually located farthest from convenient exits. Therefore locate your smoke detector in areas your family must pass through to escape. This will help prevent your family from being trapped by dense smoke or flames.

3. Locate your smoke detector on the ceiling or high on a wall. The preferred location may be in the center of the

ceiling at the top of a stairway, or at least 4 inches away from any wall. A wall-mounted detector should be from 4 to 12 inches from the ceiling. Stick to the locations recommended by the manufacturer.

4. Always locate your smoke detector at the highest point on any sloped ceiling.

5. Test your location before final installation. Again, your smoke detector has to waken all sleeping persons, even behind closed doors. Before final installation have all members of the family go to their bedrooms and close the doors. Test the detector. Every member of the family should be able to hear the alarm loud and clear.

Remember:

1. Don't install a ceiling-mounted smoke detector within 4 inches of a wall or corner.

2. Don't install your smoke detector in front of air registers. Your location should be away from windows or doors that create high velocity air drafts that might affect detector sensitivity.

3. Don't leave your smoke detector in an unheated home where temperatures may fall outside performance levels (40° to 100° F).

4. Always heed the instructions in the installation guide. Certain smoke detectors may be adversely affected by humidity, temperature extremes or stagnant air.

Since homes vary greatly in design, it's a good idea to ask the local fire department for advice. A member of the department will be glad to advise you in placement of your detector.

Make the most of your good investment.

As more and more smoke detectors enter the market, the consumer has many choices—in price, size, and style, battery or house current models.

Regardless of your choice, you'll have to keep the detector in working order. You must periodically perform a "smoke test." This may mean blowing smoke into the alarm. Some models may even have a built-in "smoke test." Read the owner's manual carefully to learn how to test your detector and how often.

You must also regularly check the power source for your smoke detector. To meet laboratory approval, all battery-operated smoke detectors must produce a distinct sound to alert you to weakening batteries. Some battery-operated models have a warning light or flag to indicate that the batteries are wearing down. As an extra precaution, always check the batteries or electrical connection when you've returned from traveling.

A detector with a dead battery or disconnected cord has no safety value and provides a false sense of security.

Finally, be sure that your smoke detector is unobstructed at all times. *Never* paint your smoke detector. Some of the paint is sure to get inside the unit, no matter how careful you are, and this will impair the unit's ability to give you advance warning in case of fire. To keep the unit free from dust and dirt, clean it at least once each year by holding the nozzle of a vacuum cleaner to the side slot openings.

Don't be lulled into a false sense of security just because you've bought and installed a smoke detector.

Even after it warns you of a fire, you have only moments to escape. Unless you have planned and practiced a family escape plan, those few moments warning might not be enough.

CHAPTER 3
Meet EDITH.

The smoke detector can give you extra time; it's up to you to make the best use of it.

Just what would you do if you awaken to the sound of your smoke detector? What would your children do?

Unless you've taken the time to sit down and plan what to do if fire strikes your home, you might not do the right thing. Each family member *must* know where to go and how to behave in case of fire.

Escape may not be as easy as walking down the stairs and out the front door.

You have to plan for the worst possible conditions; your life could depend on it.

The basis for a fire escape plan is simple: Get out! Minutes, even seconds, count. If you have a plan, you won't waste those few precious moments in trying to figure out how you'll escape; you'll already know.

Practice your escape plan: make it a habit. If the family has discussed and practiced the fire escape plan, chances are that training, rather than panic, will determine personal reaction.

Always sleep with bedroom doors closed. They are your best barrier between you and the fire.

If you awaken to smoke or the sound of your smoke detector, go to your door. Feel the door; if it's hot, don't open it. You'll have to get out another way. The hot door signals that it's too dangerous on the other side.

If the door is cool, brace your shoulder against it and open *slowly*. If smoke and heat come in, slam the door shut. Take another exit. However, if the hallway is clear, proceed carefully; your usual exit may be safe.

Now we'd like you to meet EDITH.

EDITH stands for Exit Drills In The Home. With EDITH's help, you can efficiently and quickly formulate your family's escape plan.

1. Make a list of all possible exits from your home. *You can't know in advance where a fire will start or how it will spread. It's important to have several escape routes should the first or even second exit be blocked by fire.*

_____
_____
_____
_____
_____

Now, double check every exit to make sure it really can be used in an emergency. Can children unlock and open windows, including storm windows and screens? Are hallways and escape routes clear of clutter?

2. Locate two exits from each bedroom. *Select a window in each bedroom as an alternate escape route. A*

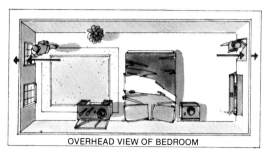
OVERHEAD VIEW OF BEDROOM

window that is painted shut or is blocked by a screwed-on screen is not an exit. Either make changes or choose another exit. Latched screens or sliding screens are the safest and easiest to operate. Be sure that everybody can reach and operate all latches, locks, doors, bolts and chains. Be sure that you can get out by every escape route you plan.

Being sure means more than knowing a window can be opened. Will you fit through? If the window is stuck and no other route is possible, is there an object nearby, such as a chair or a baseball bat, that you could use to safely break the glass?

Be sure your children know that breaking through a window is a proper and acceptable method of escape. Caution them to remove sharp pieces of glass from the window frame and to cover the sill with a rug or blanket before crawling out.

Stop the discussion now to demonstrate how to use a chair or bat, by standing to the side of the window to avoid flying glass shards, and swinging the chair or bat toward the window. Have the children practice opening the windows, too.

For easier exit from upstairs windows, buy a rope or metal escape ladder. Some escape ladders can be permanently bolted to the bedroom wall. Others hook over the window sash. Everybody should practice using the escape ladder.

3. Draw a floor plan of your home. *Even if you built your home and know every nook and cranny, you should use the grid sheet in the Firesafety Kit (in the back of the book) to draw a special floor plan of your home. The floor plan will alert you to all possible fire exits at a glance* *and will provide information to visitors if you hang it on a bulletin board or on a closet door. This is especially important for babysitters who need to know the plan, the various escape routes and the fire department phone number.*

4. Include all windows, doors and outdoor features in your floor plan. *Remember to mark all windows and doors and every outdoor feature and possible obstacle. It's important to know if you can climb out on a roof, or balcony or garage. Perhaps there is a tree that can be reached from a window. Be sure that any of these outside aids to escape can really be reached and can support the weight of a child or an adult.*

5. Indicate primary and alternate exits from every room on your floor plan. *Usually, the door will be the primary exit from a bedroom; a window will be the alternate choice. Each person must understand that if the first exit seems dangerous, he must immediately head for the second exit. Talk through and demonstrate testing a door to see if the way is clear (from page 15).*

6. Designate a meeting place outside and mark it on your floor plan. *You must know as soon as possible who is still inside. If one of your family is trapped in a room, it is vital that the fire department knows where to look first.*

Choose a spot that everyone will remember, such as a particular tree or the mailbox in the front yard. Use the front of the house if you can. That's where the fire department will arrive.

Really hammer this home—once out of the house and at the meeting place, *no one re-enters the burning house.* Leave that to the fire fighters.

7. Locate a fire alarm box or neighbor's house for calling the fire department. *Include this on the floor plan, too. Tape the fire department telephone number to the telephone and write it down on the floor plan.*

8. Go over the entire plan with every member of your family. *Discuss the floor plan and explain the reason for the choice of exit. Walk through the escape plan, taking the family through the escape routes for every room. Use this walk-through to double-check your escape routes.*

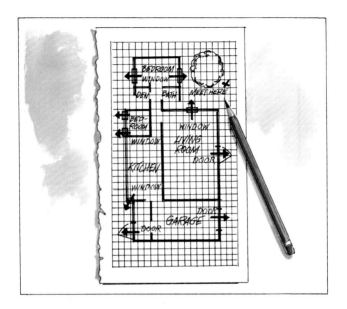

Have your children practice saying the fire department phone number, the family name, street address and town into the phone. Make sure they understand that they may have to run to a neighbor's house to make this call.

9. Train every child in your home to follow the plan.
Children panic in fire. Often their bodies are found under a bed, in a closet, behind furniture; places they had gone to hide.

Teach your children that they can't hide from fire; they must escape it. You can't be too firm on this point.

The better a child has been taught the escape plan and the more times he or she has taken part in the fire drill, the better the chance of correct reaction to a real fire.

You may decide that the best action is for the child to stay in his room until you arrive to get him. But what if you become trapped or find the path to the child's room blocked?

As early as possible, the child must be taught how to escape alone.

10. Go over the entire plan with a representative of your local fire department, informing him of any special circumstances. *In most towns and cities, you can do this in your own home while the fire department representative makes a simple, free firesafety check.*

Get his advice on your escape plan, and change it if he suggests. Be sure to point out an infant's room or the room of an elderly or handicapped person.

Mount your house number conspicuously in large, readable numbers against a contrasting background so the fire department can quickly find your house.

11. Conduct a fire drill at least once every six months. *This is a serious business: the business of your family's safety. After installing an alarm and discussing how to behave in case of fire, be sure that your family can remember and follow the plan.*

In a real fire you must be prepared to move rapidly and carefully, without panic. You need to make the drill as realistic as possible, but don't rush through it as fast as you can. The point is to make sure that everyone knows what to do, not to win a race.

As your family works out its escape routine, vary the drill by calling out different hazards and different fire sources. For example, one drill might place a supposed fire in the kitchen, while another might place it in the family room. Be sure that everyone understands how the differences in the fire's location might alter escape routes.

At the beginning, you might stay back and coach your children, being sure that they check for the hazards that you describe to them during the course of a drill.

Don't warn them of the fire drill in advance.

Practice is important, but it's not the only reason to conduct your fire drills on a regular basis. The drill keeps your escape plan up to date and realistic for all family members. An escape route that was impossible for a four-year-old six months ago may be easily handled by the child today.

Never hesitate to change your plan as needs and abilities change. After the drill, gather the whole family to discuss what took place and how to improve on performance. You might decide to have another drill in two or three weeks, rather than wait six months.

CHAPTER 4
Survival

Even though your smoke detector should provide an early fire warning, and your escape plan a simple escape routine, you must be ready for the unexpected.

In an actual fire, your survival may depend upon how well you and your family have learned the escape plan and the dangers to look for. You must be prepared to react without panic in a calm and deliberate manner. It isn't easy.

You have just awakened to the sound of your smoke detector. You are still groggy. Perhaps you are already inhaling small quantities of toxic gases from a fire downstairs.

You know the escape plan. You may need to use some additional survival tips. Your escape plan and behavior *help you* accomplish the same purpose: getting out.

Explain these survival tips as the family discusses EDITH.

1. Remember, sleep with the bedroom doors closed. *The closed door offers protection from heat and smoke. Even a lightweight hollow core door may provide you with extra time to escape by slowing a fire's progress. As you proceed through your escape route, you should close every door behind you.*

2. Be sure everyone recognizes the sound of the smoke detector and knows that it means GET OUT NOW! *If you've had a fire drill, this is the moment to set*

your escape plan in motion. If you haven't yet held a drill, you must be certain that all family members recognize the sound of the smoke detector. Just in case, call a warning, such as "FIRE! EVERYONE OUTSIDE!"

3. Never waste time getting dressed or gathering valuables. *You don't have time to waste. You can borrow clothing later. No valuables are worth risking your life and the lives of your family.*

4. Feel every door before you open it. *If the door is hot or if smoke is coming through the bottom or sides of the door, don't open it. Get out by an alternate exit.*

Even if the door is cool, you must be careful when you open it. If the door is cool and there is no smoke, brace yourself against the door, turn your face away and open it carefully. Be ready to slam the door shut if the slightest trace of heat or smoke enters the room. A fire that has died down for lack of oxygen could flare up once the door is open. If any smoke or heat does enter the room, slam the door and latch it tight.

If nothing happens when you open the door, go ahead and use your primary escape route. *Remember to close every door behind you.* This keeps the fire from spreading.

5. Use windows for escape and rescue. *If your first exit, the door, is unsafe, you will have to use the second choice—probably a window—to escape the fire. To aid in using windows, you may need escape ladders for use in upper story bedrooms.*

Your family needs to know the following steps in using the window to escape a fire.

First, make certain the door is closed tightly. Otherwise, the draft from the open window may draw smoke and fire into the room.

If the window is more than two stories from the ground and you don't have an escape ladder, balconies or other structural aids to use for escape, you should not jump, but wait for the fire department. In such a case, open the window a few inches at both the top and the bottom. Fresh air can enter from the lower opening and any smoke or gases can ventilate through the upper portion. Stay at the window, keep your head low to get fresh air, and wave a light-colored towel, sheet or some other obvious signal to help the fire fighters locate you.

If on the first or second floor, you can probably drop from your window.

It may be helpful to move a chair or some other piece of furniture to the window to make it easier to climb out. Either straddle the window sill or back out of the window, maneuvering so you can slide out on your stomach, feet first. Hold on with both hands to the sill. Now let go of the sill and drop to the ground, bending your knees to cushion the landing.

You can lower small children from the window. There may be someone outside to catch the child and help break the fall. If not, you must lower the child as far as possible. Don't leave the room first and expect young children to follow you. They could panic and you would have no way of helping them out if you were on the ground below.

6. If there's smoke, crawl: Get down on your hands and knees. *Do not stand up.*

In a serious fire, superheated air will rise to the ceiling. It may have a temperature of 1,000° F up there. Keep down. Below the superheated air, but still near the ceiling, is a thick layer of carbon monoxide—a tasteless, odorless and colorless gas. Stay down! Smoke—the deadliest killer produced in a fire—will remain at the same level as the carbon monoxide. If you see smoke, crawl. Get down on your hands and knees, not your belly. Some of the toxic gases are heavier than air and will also settle in a thin layer on the floor.

Crawling on your hands and knees is safest because it keeps you in a roughly defined zone of safety that exists between 12 inches above the floor and doorknob height.

7. Drop and roll. *If your clothes catch on fire, drop to the floor and roll to smother the flames.*

Be certain that all children and adults in your home know *never* to run if their clothes catch on fire. Running just gives the fire more oxygen.

As part of the family practice, stage a "drop and roll" session. Also tell your family how to help if someone else's clothes catch on fire. Tackle or knock the person down and make him roll. Or, get him to the ground and throw a blanket or rug around him to smother the fire.

8. Once you are out of the house, go immediately to your meeting place. *You have to find out who is still inside so the fire fighters can be informed when they arrive.*

9. Call the fire department. Make sure you give your complete address and say if you think someone is trapped in the fire. *Don't attempt to call from your home. It's too dangerous.*

Call from a neighbor's house. Fire departments need to know the address first, and information about people trapped inside.

Try to be calm when you call the fire department. Give them whatever information they ask for and be sure to stay on the telephone until you have answered all questions. The fire fighters will be on their way immediately as you talk to the dispatcher. A longer call will not delay their arrival.

If there is a fire alarm box nearby, use it and stay there until the fire fighters arrive.

10. Once you are out, stay out. Never return to a burning building. *The professional fire fighter is trained in saving lives as well as in putting out fires. His specialized training, protective clothing and emergency air supply give him an edge.*

11. Don't permit anyone to go back into the building. *Just as you may have to fight the urge to return to your home and try to save trapped family members, others may be fighting the same urge. Do not permit anyone to go back in; keep a firm hand on small children. The risk is too high.*

12. Cool a burn. *If someone gets burned, don't gamble. Get to a doctor or hospital immediately. When a burn happens, use cold tap water to stop the heat injury. Don't use ice or butter or ointments. Cool the burn with water immediately. Delay worsens the injury.*

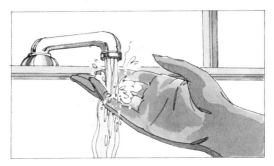

Remember that fire and smoke are a deadly combination that requires calm, deliberate action. With a clear understanding of the nature of fire and how it functions, you and your family should survive if fire strikes.

With an early fire warning system and escape plan, you have both peace of mind and the key to survival in case of a fire.

CHAPTER 5
Specially Tailored Plans

The handicapped and elderly

Elderly and physically handicapped persons should have a separate smoke detector in their rooms and a separate plan tailored to their special needs. Be sure that the fire department knows that an elderly or handicapped person lives in the house. (Mention this when you go over your escape plan with a fire department representative.)

If possible, the person's room should be on the ground floor with a door leading directly to the outside. If the bedroom must be on the second floor, work out a special escape plan. If the person is capable of using an escape ladder, buy one for his/her bedroom. Since an elderly or handicapped person might need extra time to escape or call for help, install a telephone in that person's bedroom with the fire department number attached.

Young children

All people are fascinated by fire. As adults, we can control our fascination with the knowledge of fire's potential danger. Until taught or until suffering a painful experience, a child does not understand. That can make a child a fire hazard, not only to himself, but to all of us.

Children under the age of four or five may not be able to fend for themselves in a fire. If possible, young children's bedrooms should be close to the parents' room or to an older brother's or sister's room, making it as easy as possible to bring the children out quickly. Where it's practical you might consider placing connecting doors between the rooms.

The biggest service you can do for very young children is to make firesafe living a major part of their daily routine. Certainly the learning can be fun for them. But make sure they *learn*.

Make sure children learn the basics of firesafety. Include them in your escape planning and fire prevention inspections (see Chapter 6). Be sure that they know not to play with matches and other heat sources.

Remember, too, that hot liquids can burn like fire. Keep pot handles turned in and electric coffee pots out of children's reach. A curious child could tip the pot, pouring boiling liquid on himself.

Test the bathwater—and teach your children to do the same—before bathing a child. Even ordinary tap water can cause second and third degree burns.

You may not be able to reach your children in a fire. They must understand how to get out of their rooms by themselves. During EDITH and your family discussions on escaping fire, emphasize your confidence in their ability to get out.

Apartments

If you live in an apartment, you should plan to follow the Survival Points listed in Chapter 4 whenever possible, but there are some variations.

Plan your escape with EDITH, making sure to locate the nearest alarm box during the planning stages. The alarm will warn the other tenants in your building of the fire. If there is no alarm box, warn your neighbors by pounding on other apartment doors as you leave. Close all doors behind you. Let the fire department evacuate the rest of the residents.

Never use an elevator. It could open on the floor of the fire where heat and smoke could kill you.

If your apartment is two stories or more above the street level, do not jump from your window to escape. Here, too, an escape ladder can be a great aid.

If fire blocks your exits, close your apartment door. Cover all the cracks where smoke may enter the room. Phone the fire department, even if they're on the scene, and tell them where you are trapped. Open a window at the top and bottom; fresh air enters from the bottom, smoke will flow out the top. Wave a length of sheet or towel. The fire fighters will see your signal and will get you out.

Babysitters

When you leave your children in the care of a babysitter, take time to cue the babysitter to your family's escape plan. Discuss the floor plan with marked exits and point out the telephone number of the fire department.

Use the Babysitter Checklist in the Home Firesafety Kit as a quick reference for the babysitter. Also leave the phone number where you can be reached.

Explain the family firesafety plans to your babysitter. Make it clear that your only concern is that everyone gets out of the house quickly and safely. If the alarm sounds, no one should try to find the fire, much less try to put it out.

Mobile homes

If you live in a mobile home, follow the same procedures that apply to conventional dwellings. In addition, you should be especially careful of gas-fired appliances. Be sure that gas heaters are vented to the outside, and that LP gas tanks are *never* stored inside your home. Keep all exits clear and easily accessible.

Pets

If you own pets, you have undoubtedly noted that there was no mention of household pets in the chapter on escape planning. This was a deliberate omission. The dangers of a serious fire are so overwhelming that the primary concern is saving human lives. You simply do not have time to consider anything else.

In any case, your larger pets—dogs and cats—are as terrified of the fire as you are. Often, they will escape on their own, even before you do.

If all the people in your home escape and gather at your prearranged meeting place, it is likely that your pet will join you. If the pet remains trapped inside, tell the fire fighters where they might find it. Fire fighters will often try to save pets from fires. *You* should *never* try it.

Don't even waste time bringing a pet along with you as you escape. Make sure your children understand the importance of not going back into the house under any circumstances—and keep an eye on your children to be sure they *don't*.

The necessity of leaving the family pet inside a burning building is difficult to explain to a child. But the alternative is worse. Be firm and explain your decision as fully as the child can understand.

CHAPTER 6
Learn Not To Burn

A smoke detector, an escape plan, provisions for special situations—none are fail-safe guarantees.

The only real safety from fire is keeping fire from happening in the first place.

You and your family must Learn Not To Burn. As discussed in Chapter 1, fire needs oxygen, fuel and heat in order to burn. Understanding the major causes of home fires may help you to keep these elements apart.

1. Carelessness with cigarettes, cigars and pipes. *This is the single largest cause of home fire. Always use an ashtray when smoking and don't permit ashtrays to become too full; hot ashes may fall over the edge. Use large, heavy ashtrays that won't tip over. When you empty an ashtray, make sure nothing is burning.*

<u>*Never smoke in bed*</u>*. A smoldering mattress can kill you with smoke and poison gases long before flames appear. You may never wake up.*

Before going to bed after parties, check under cushions and behind furniture for smoldering cigarettes. A chair with a burning cigarette under its cushion might smolder for hours before bursting into flames.

2. Faulty electrical wiring. *Faulty and misused electrical wiring is the second leading cause of home fires.*

Check with an electrician to be sure that electrical circuits aren't overloaded.

Make certain that your home has enough electrical circuits to avoid overloading. See that special appliances, such as air conditioners and large space heaters, have their own heavy duty electrical circuit.

Always replace blown fuses with a fuse of the proper size; never use a penny.

An overloaded extension cord presents a fire hazard. Too many appliances plugged into an extension cord can ignite the cord's insulation.

Consider this investment for firesafety: have additional wall outlets installed. This way, you won't have to rely on extension cords for electrical appliances.

Don't run electrical or extension cords under rugs or carpets and don't hang cords on nails. Eventually, the cord insulation may deteriorate, exposing a live wire.

The sparks and heat of a short circuit could be the ignition source of a fire in your home. Take steps not to let that happen.

3. Faulty lighting equipment. *Check electrical cords for cracks, broken plugs and poor connections. Lamps that fall over easily are potential hazards, as are lampshades that touch or are close to bulbs. And that marvelous old lamp you found at the neighborhood garage sale should be taken apart and rewired before use. Use the proper size light bulb with lamps and lighting fixtures.*

Be sure that all electrical appliances are in good condition at all times.

4. Carelessness with cooking and heating appliances. *If you have a coal or wood heating stove, check with the fire department to make sure that the stove is installed* *properly at a safe distance from combustibles. Cover the fireplace with a metal screen to keep hot sparks from dropping into the room.*

Have your heating system professionally inspected once each year.

As a day-in, day-out precaution, be alert to cooking habits. Keep pot handles turned in from the stove's edge. Don't leave food cooking unattended on the stove. Don't store food above the stove where a child, or grown-up, may be burned while reaching over the hot stove. Always

wear short or tight-fitting sleeves while cooking; loose clothing could catch fire.

In case of a grease or pan fire, smother the flames with a pot lid, a larger pan, or use a portable fire extinguisher. <u>Never</u> throw water on a grease fire. In case of an oven fire, close the oven door and turn off the oven.

5. Children playing with matches. *There are two things you must do. First, make sure your children understand the danger of playing with matches, lighters and other* *ignition sources. You may be surprised to find how much a young child can be made to understand. Second, keep matches and other ignition sources out of the "strike zone"—that area from the floor to the shoulders where small children can reach. Teach older children how to use matches safely.*

6. Open flames or sparks. *If you use candles or oil-burning lamps, make sure they and their holders are in good condition before each use and that they stand securely. Never leave candles burning when you leave a room for more than a few moments.*

Buy flashlights for use in emergencies—don't use candles. Never use an outdoor barbecue indoors, not only because of the fire hazard, but because of the toxic gases and vapors produced by smoldering charcoal. When you use the barbecue outdoors, be sure to keep it a good distance from the house and never leave the fire unattended for more than a few moments. Be sure that the coals are cool before you leave the barbecue.

7. Flammable liquids. *All flammable liquids are dangerous. At temperatures below zero, a single spark can ignite gasoline vapors. Even ordinary fingernail polish remover is a flammable liquid.*

Don't take chances. Never smoke when you work with flammable liquids. Don't store or use them near any heat or ignition source.

If you must keep a small amount of gasoline on hand to power a lawnmower or similar equipment, don't store the gasoline in the house. Keep it in a garage or tool shed. Use a safety can—one with a spring closure valve, vapor vent, pouring spout, and the label of a testing laboratory. Never store any flammable liquid in a glass jug, discarded bleach bottle, or other makeshift container.

Remember that many spray can products, such as paints and hair spray, are combustible liquids held under pressure. Use these carefully and follow the directions on the label.

8. Faulty chimneys and vents. *Have chimneys and fireplaces inspected yearly for cracks, crumbling bricks, obstructions and creosote accumulations. If inspection uncovers problems, contact a chimney sweep or repair company to clean or to make necessary repairs.*

Permanently seal off unused flue openings. Don't use rags or snap-in flue covers.

Check gas vents for corrosion and obstructions that could present fire hazards.

9. Arson. *Arson is the seventh ranked cause of residential fires. If someone really wants to set fire to your home, there is little you can do to prevent it. But you can make it more difficult. Most fires are set from the outside of the home. A pile of trash or a can of gasoline left in plain sight outdoors not only makes the arsonist's job easier, but might even attract him.*

Fire—an invading enemy.

If you approach the problem as a defending General, your job becomes clear. To exist, fire needs oxygen, fuel and a heat source. Oxygen cannot be removed from the home, so we must attack the problems presented by fuel and heat. Most anything can become fuel for a fire. But some materials burn more easily than others.

Wherever possible, you should be sure that your home is finished with and furnished with fire resistant or fire retardant materials. This includes the roofing, as well as the interior surfaces and furnishings of your house.

If you are building or remodeling your house, keep firesafety in mind when buying building materials. Your choice of products can contribute to the safety of your family.

Walls — *Buy gypsum or other noncombustible wall boards. These offer considerable fire resistance. Dry wall (gypsum) or plaster are the most commonly used for firesafety reasons.*

Paneling — *Wall paneling may contribute to fire spread, unless you buy paneling labeled as fire retardant. Look for a label that gives the flame spread rating of the paneling. In new constructions, have a layer of ½" gypsum board put between paneling and insulation.*

Ceiling — *Most ceiling tiles made today have reduced flame spread characteristics. Read labels carefully and buy only ceiling tiles that have the label of a testing laboratory.*

Insulation — *Insulation can really pay off in reduced heating and cooling costs. Before installing it, check your home's electrical system for any problems. When purchasing the insulation, check for the seal of a testing laboratory and for flame spread ratings.*

Be sure not to pile insulation around recessed lighting fixtures; maintain a minimum 3-inch clearance to avoid heat build-up.

Alternative heating — *Factory built fireplaces and wood stoves can help offset rising energy costs, if you are careful and firesafe in selection, installation and maintenance.*

When buying a wood stove, look for: sturdy construction of cast iron or heavy steel; label of a testing laboratory; danger signs of cracks or punctures; a damper to control the draft.

Follow required clearances—minimum safe distances between stove and the walls and ceiling—as given in manufacturer's instruction and local safety codes. Minimum clearance is 3 feet, but may vary depending on the stove and protective covering on walls.

Usually, 24-gauge sheet metal or 4-inch hollow tile is required as floor covering beneath the stove.

Keep combustibles such as clothes, furniture and newspapers away from stoves and fireplaces. Never burn trash.

Burn well-seasoned wood. Underseasoned or green wood burns inefficiently and unburned gases form creosote on chimney walls. If ignited, creosote creates a very hot chimney fire. Chimney fires can also be caused by overfiring—letting a fire burn too big.

Allow ashes to cool before disposing of them in a tightly-covered metal container; never use boxes or bags.

Special care should be taken with plastics, keeping them away from heat. Most household plastics do not ignite easily. But some plastics emit large volumes of extremely toxic gases when they burn or smolder.

Fabrics offer special concern that you should consider. Most will burn with relative ease. And when clothing burns, you burn. Fuzzy, lightweight, loosely woven or loose fitting fabrics ignite and burn easily. Look for sturdy fabrics with a smooth, tight weave. Denim and wool are less likely to burn quickly. Sleepwear for children up to 12 years must be labeled with fire retardant information. Pay attention to the laundry instructions on fire retardant clothes, otherwise you lose the protection.

UNSAFE—FUZZY, LIGHTWEIGHT, LOOSE FABRIC

SAFE—STURDY, TIGHTLY WOVEN FABRIC

Finally, use the Fire Prevention Checklist in the Home Firesafety Kit and be sure that your family is familiar with common fire hazards and on the watch for their occurrence.

Make a scheduled fire prevention inspection tour of your home at least twice a year. Correct any hazards you find *immediately.* It lessens the chance of fire and helps to slow the progress of a fire if one does break out. Remember—fire prevention is your best defense against fire.

CHAPTER 7
Firesafe Life Style

Tour your home with firesafety in mind. Does your home measure up—from basement to attic—from room to room? You may be nodding yes, but are you sure?

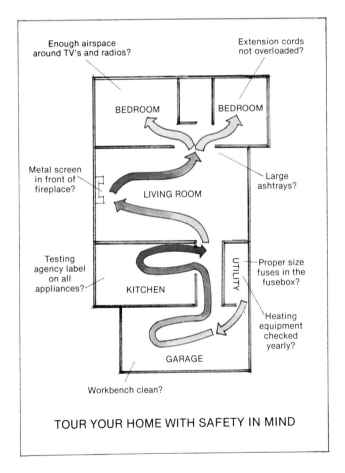

- *Start in your basement or storage room. Fuse box equipped with proper fuse sizes? Heating equipment checked yearly? Rubbish dumped regularly? No combustibles stored near the furnace or heater?*

- *Garage—gasoline stored in a safety can? Never store in the garage or basement where there's a gas hot water heater. The gas pilot light can ignite vapors from flammable liquids.*

As for hobbies and work in these areas, does everyone follow the firesafe guidelines of keeping flammable liquids away from heat, pilot lights and sparks? Make sure the workbench and floor are clean and clear of wood shavings.

- *In the kitchen, check to see that each electrical appliance has a testing laboratory label. If an appliance isn't working right, don't use it until it has been repaired. Don't overload extension cords; this presents a serious fire hazard.*

Remember not to store food over the stove. In reaching for the food, someone could get burned or clothing could catch fire from a hot burner. By the same token, remind all cooks in the family to wear short or tight-fitting sleeves when cooking. Loose clothing can catch on pot handles, or worse, can catch on fire. Keep pot handles in, to avoid accidental spills and scalds.

- *Move on to the living and family room. Are large ashtrays available for smokers? Do you have a sturdy metal screen or heat-tempered glass doors in front of the fireplace? Allow air space around televisions and stereos to prevent overheating. Locate portable heaters away from doorways, combustibles and the usual traffic paths of your family.*

- *Check bedrooms. Enough air space around appliances? Large ashtrays available for smoking in a chair, not in bed? Again, extension cords not overloaded?*

In every room of your home, make certain that matches are out of the "strike zone," well out of children's reach.

Special occasions.

Camping. When camping out, never have a flame in the tent. This includes oil lamps, candles, matches, heaters of stoves. Before buying a tent, make sure it is labeled "fire retardant."

Have an adult be responsible for matches; keep matches and lighters away from children.

Barbecue. Never use gasoline on an outdoor barbecue fire. Once the fire has started, don't use any liquid fire starter. If the fire burns too low, use dry kindling or blow gently across the charcoal to revive the flame.

Never burn a charcoal fire inside your home.

Outdoor firesafety. A single spark can ignite highly flammable gasoline vapors. Follow these suggestions if you use gasoline-driven appliances, such as motor bike, lawnmower,

CAUTION: GASOLINE IS HIGHLY FLAMMABLE AND DANGEROUS

chainsaw, snow blower, garden tractor: After filling a gasoline tank, move the machine away from the gasoline fumes before starting the motor. Always allow the motor to cool before refueling. Never use gasoline to get paint off skin or clothes or to clean equipment parts.

If you must carry gasoline from service station to home, use an airtight, unvented can filled only three quarters full.

Holidays. <u>Fourth of July</u> — This holiday wouldn't be the same without colorful public fireworks displays, but only professionals should use fireworks.

• *Fireworks are unpredictable. Some fire prematurely. Apparent duds will explode without warning. Even non-explosive fireworks such as sparklers may ignite clothing.*

• *Don't use fireworks; steer clear of people who do. Don't let tragedy mar your July Fourth: Let the pros handle fireworks.*

Halloween — Protect your family ghosts and goblins by following these simple suggestions:

- *Don't use flimsy materials and voluminous amounts of old sheets for costumes. Also beware of paper-bag masks and paper costumes—these can ignite in a second.*
- *Ready-made costumes, masks and wigs should be labeled "flameproof." Don't buy them if they're not.*
- *Never use lighted candles in hand-carried pumpkins. Substitute a small flashlight.*
- *Be extra careful with harvest season decorations. Keep them away from fireplaces and other sources of heat, and don't let them block doorways or stairs.*

Christmas checklist — Ironically, there's an increase in home fires and related deaths during the traditionally joyous Christmas season—due to the age-old problem of carelessness and to the built-in fire hazards associated with Christmas decorations. Don't let your family fall victim to a holiday fire. Read through this Christmas checklist and see to it that your holiday is free from fire.

Lighting safety.

- *Lighting sets must be labeled by a testing laboratory.*
- *Replace worn sets.*
- *Never use electric lights on a metal tree.*
- *Don't overload extension cords. Protect wires from injury—don't run wires under rugs.*
- *Outdoors, use only lighting specifically made for outdoor use.*
- *Don't burn real candles in the windows.*

- *Unplug all lighting before you go to bed or leave the house.*

Children's safety.
- *Don't leave your children alone for a minute or you invite accidents, especially at Christmas time.*
- *If you buy plug-in electric toys, be sure that they have been labeled for firesafety by a testing laboratory.*
- *Don't buy highly combustible toys, or any that use flammable liquids.*

Your tree.
- *Don't buy a tree with shedding needles.*
- *Cut off one inch from the trunk. Keep the tree in water in a non-tip stand, away from exits and sources of heat.*
- *Don't use real candles on the tree.*
- *Plastic trees should have a fire-retardant label.*

Parties. Candles may make nice decorations—they also cause fires. If you use candles for party decorations, be careful where you put them—never near curtains, doorways, plastic or paper decorations. Your best bet is to keep the candles in hurricane lamps. Use flame retardant or noncombustible decorations and costumes.

When the party ends and you're making the clean-up rounds, make a special check for smoldering cigarettes. Look under cushions, behind furniture, in waste baskets—any spot that could conceal a hot cigarette butt.

CHAPTER 8
When You're On The Spot

Using fire extinguishers.

There may be a time when you see a fire start and are close enough to put it out. In such a case, having a fire extinguisher may keep that small fire from becoming a large one.

The only times you should try to extinguish a fire by yourself are when you're near the fire when it begins or you discover the fire in its early stages, and you know how to use the extinguisher. Have someone call the fire department, even as you try to extinguish the fire.

Just remember, your first concern is getting the family out and calling the fire department.

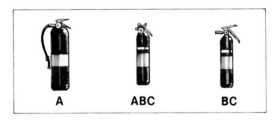

There are several classes and sizes of fire extinguishers suited for home use. Extinguishers are rated for use on Class A fires involving ordinary combustibles such as paper or wood, Class B fires involving flammable or combustible liquids, or Class C fires involving electrical equipment. Extinguishers for all three classes are available in one- to five-pound capacities or larger.

If your home has only one extinguisher, your best choice is a multi-purpose (A:B:C) dry chemical extinguisher, which applies to all three classes of fire.

The label will specify the purpose of the extinguisher and give you instructions for its use.

You should have at least one extinguisher located in the kitchen. It's a good idea to have an extinguisher in your workshop, too. Locate each extinguisher in the path of exit travel so that there's an escape route if the fire cannot be controlled.

These extinguishers have a discharge time of only eight to twelve seconds. Obviously, it's important that family members know how to use the extinguisher quickly.

As part of family firesafety planning, make a point to explain and demonstrate (outdoors) how to hold and activate the extinguishers, following the operating instructions imprinted on the extinguisher. Just go through the motions; don't discharge the extinguisher during practice sessions.

Inspect each extinguisher at least once a month. Check the pressure gauge and lift the extinguisher off the hook or bracket to make sure it's in good physical shape and easy to remove.

A thorough examination for repairs or recharging should be done by a competent recharging service company after each use or when an inspection pinpoints a need.

An ordinary garden hose can also aid in fighting fires in the home. Connected to the home water supply, the water hose provides unlimited discharge. This is especially important in outlying areas where considerable time may be required for the fire department to arrive at the scene of a fire.

Any garden hose, at least ½-inch in diameter and equipped with an adjustable nozzle, would be a good choice. The hose should be long enough to extend to potential fire areas. To conveniently install the hose under a sink or in a closet, use the hose reels available for garden use.

To be effective, the hand hose should remain connected to the domestic water supply, be regularly checked for deterioration and be used only in fire emergencies.

Follow the same basic steps for extinguishing a fire with either the garden hose or an extinguisher. Aim the water hose or extinguisher at the *base* of the fire, using a sweeping motion.

Don't get so close to the fire that you endanger yourself.

Leave the fire if you feel you may be in danger. In all cases, call or have someone else call the fire department *before* you try to fight the fire. Don't delay in alerting the fire department, or it may be too late.

A FINAL WORD

The information in this book and the time your family spends in planning for firesafety may save your life.

For further guidance, call your fire department. Fire fighters are a concerned and knowledgeable source of information. They put out fires, but they're also dedicated to preventing fires.

You can also write directly to the Public Affairs Division of the National Fire Protection Association (NFPA) at 470 Atlantic Avenue, Boston, Massachusetts, 02210.

Founded in 1896, the NFPA is the only worldwide group dedicated to the protection, through education and science, of man and his environment from fire.

The Association develops and updates firesafety codes and standards. Though advisory, many of the codes and standards have been adopted into federal, state and local law due to NFPA's reputation for accuracy and integrity.

NFPA provides technical assistance on firesafety relating to such topics as flammable liquids, nuclear energy, and life safety in buildings. The Association has the largest collection of fire information in the world.

Additionally, NFPA reaches millions of people through its public awareness campaign that includes Fire Prevention Week, Spring Clean Up and the "Learn Not To Burn" television public service announcements with Dick Van Dyke.

HOME FIRESAFETY KIT

While you are working out your family firesafety plan, use this checklist as a double check. Simply cut the sheets from the book along the dotted lines. After you've toured your home, post the checklist as a reminder to keep all areas firesafe. Chapter 7 has several tips that will help you in your home fire check.

Remember these four ways to protect your family:
1. Correct any household fire hazards you find.
2. Install smoke detectors in the proper locations.
3. Practice family escape planning.
4. Teach your family firesafe behavior.

With your family together, go over these things you can do to keep fire out of your house.

☐ Are fuel-burning space heaters and appliances properly installed and used?

☐ Has the family been cautioned not to use flammable liquids like gasoline to start or freshen a fire (or for cleaning purposes)?

☐ Is the fireplace equipped with a metal fire screen or heat-tempered glass doors?

☐ Since portable gas and oil heaters and fireplaces use up oxygen as they burn, do you provide proper ventilation when they are in use?

☐ Are all space heaters placed away from traffic? Children and old persons cautioned to keep their clothing away?

☐ Are proper clearances provided between space heaters and curtains, bedding, furniture?

☐ Do you stop members of your household from smoking in bed?

☐ Do you check up after others to see that no butts are lodged in upholstered furniture where they can smolder unseen at night?

- [ ] Do you dispose of smoking materials carefully (not in wastebaskets) and keep large, safe ashtrays wherever people smoke?

- [ ] Are matches and lighters kept away from small children?

- [ ] Are all electrical cords in the open—not run under rugs, over hooks, or through door openings? Are they checked routinely for wear?

- [ ] Is the right size fuse in each socket in the fuse box and do you replace a fuse with one the same size?

- [ ] Children get burned climbing on the stove to reach an item overhead. Do you store cookies, cereal, or other "bait" away from the stove?

- [ ] Do you keep basement, closets, garage, yard cleared of combustibles like papers, cartons, old furniture, old rags?

- [ ] Are gasoline and other flammable liquids stored in safety cans (never glass jugs, discarded bleach bottles or other makeshift containers) and away from heat, sparks, and children?

- [ ] Is paint kept in tightly closed metal containers?

- [ ] Are furnace, stove and smokepipes far enough away from combustible walls and ceilings and in good repair?

- [ ] Is your heating equipment checked yearly by a serviceman?

- [ ] Is the chimney cleaned and checked regularly?

- [ ] For safety against chimney and other sparks, is the roof covering fire retardant?

- [ ] Are there enough electrical outlets in every room, and special circuits for heavy-duty appliances such as space heaters and air conditioners?

- [ ] Do you have a qualified electrician install or extend your wiring?

- [ ] Do all your appliances carry the seal of a testing laboratory?

BABYSITTER CHECKLIST

Cut this worksheet out along the dotted line. Then fill in the blanks and keep the checklist handy for your babysitter. Take the time to tell your babysitter—and write down—where you will be and how you can be reached. More suggestions for firesafety and babysitters can be found in Chapter 5.

PARENT'S BABYSITTER CHECKLIST

Post by each phone. Your babysitter should be familiar with each item.

Emergency Phone List:

Family Name	_____
Address	_____
Town	_____
Phone No.	_____
Fire	_____
Police	_____
Medical	_____
Poison	_____
Neighbor's Name	_____
Neighbor's Address	_____
Neighbor's Phone	_____
Nearest Fire Call Box	_____

Checks for Babysitter:

- [ ] The doors are locked.
- [ ] The children are never left alone even for a minute.
- [ ] I know the dangers to children of matches, gasoline, the stove, deep water, poisons, falls.
- [ ] I know the location of all exits here (stairs, doors, fire escape, windows, other) and phones in case of emergency.

Names and Ages of Children:

ESCAPE PLAN IN CASE OF FIRE

This is a worksheet for you to use in mapping out your family escape plan. Simply cut it out and spread it flat. Follow the suggestions in Chapter 3 in drawing your home floorplan and marking the various escape routes. When you finish, post the escape plan on a bulletin board as a constant reminder. See the sample plan on page 18.

Sketch floor plan(s) of your house, indicating all possible fire exits with an arrow (be sure to show all doors, windows, front and back stairs, fire escapes, adjacent buildings or other ways out).

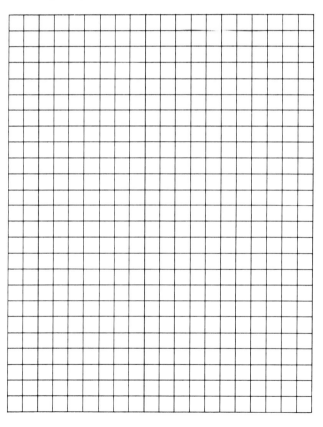

ADDITIONAL PAGE FOR SKETCHING YOUR ESCAPE PLAN ⟶

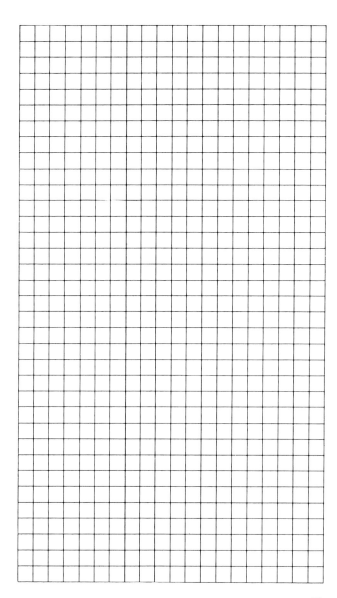

ADDITIONAL PAGE FOR SKETCHING YOUR ESCAPE PLAN ⟶

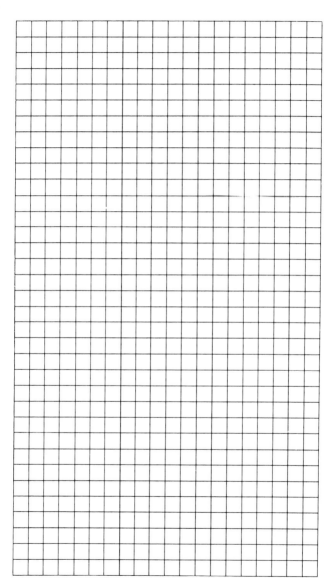